SPORTS CARS

COLORING BOOK

FOR KIDS AND ADULTS

Sports Cars Lovers!

We at Driven Hard Books love cars and we are very excited to share that love of coloring cars with you!

Here are some ways that you can get the most from our coloring books:

- All of our coloring books are designed to not only let you color but to also let all your artistic abilities shine through.
- Each car page is *your blank canvas* for you to fill with inspiration of whatever comes to mind. So let your inner artist guide the way and enjoy the ride.
- Certain pages contain grayscale shading, meant to add another layer of coloring. For the light gray areas we recommend use light colors, the medium gray areas medium colors, and the dark gray areas use dark colors.
- There is no right or wrong way to color, just relax and unwind.

We would love to see your creations so feel free to post the pictures of your coloring or your drawings on our page on Amazon.

We have made this QR code for you that will take you directly to the Amazon review page!

We can't wait to see what type of amazing designs you create!

This Book Belongs To:

Test Your Color

or if you prefer, remove this page and use it as your bleed-through protection
by placing it after the page you are coloring and moving it as you go.

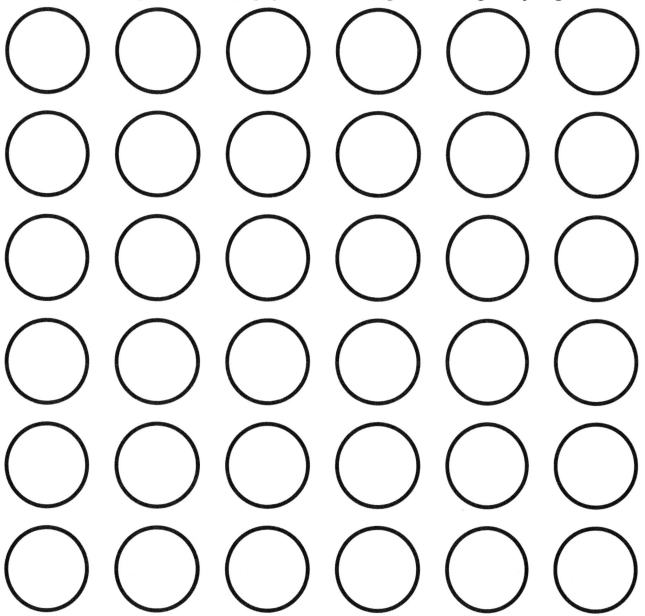

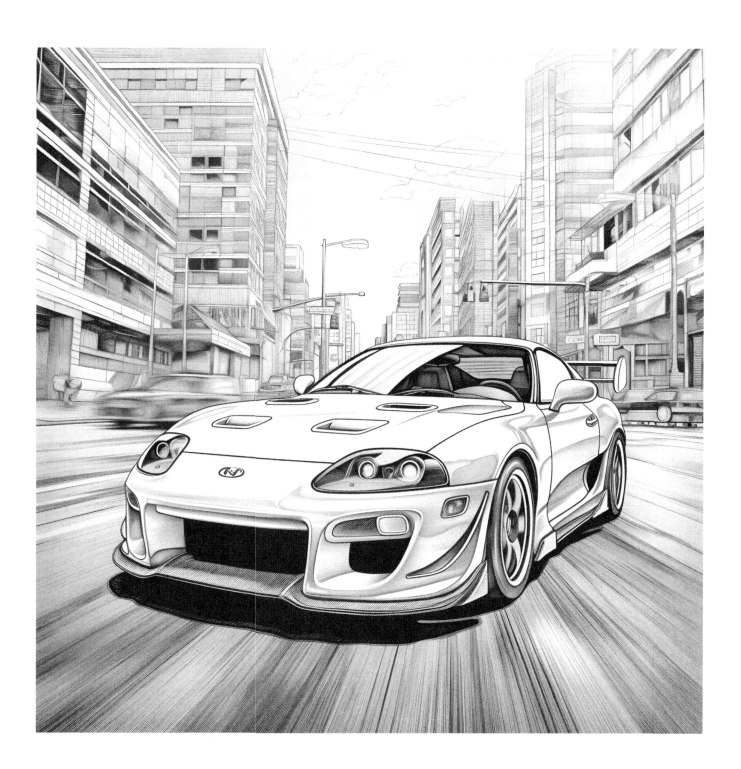

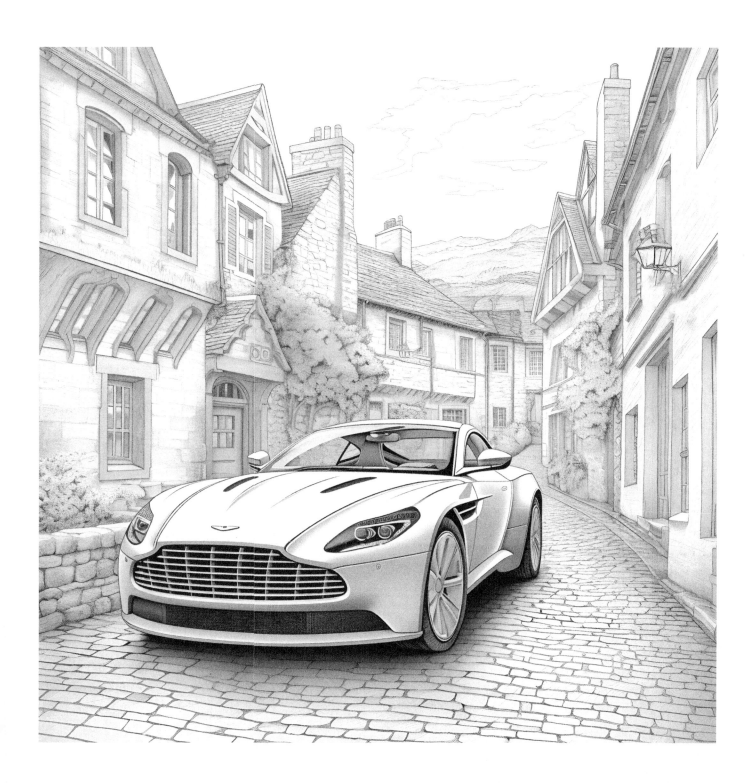

Test Your Color

or if you prefer, remove this page and use it as your bleed-through protection
by placing it after the page you are coloring and moving it as you go.

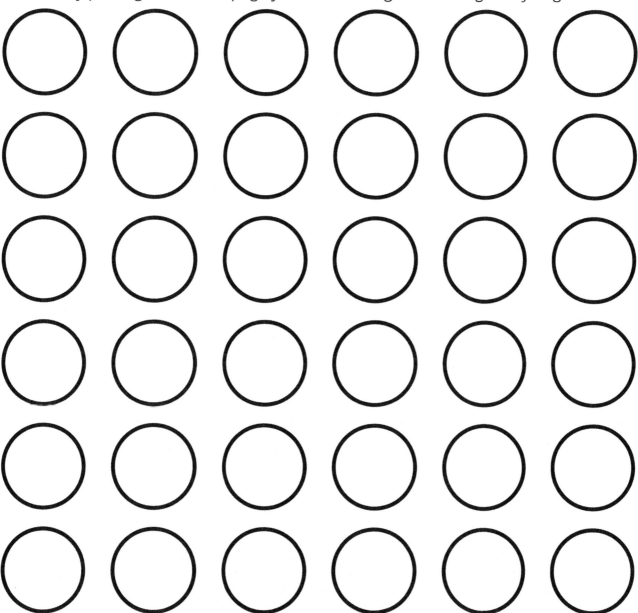

Printed in Great Britain
by Amazon